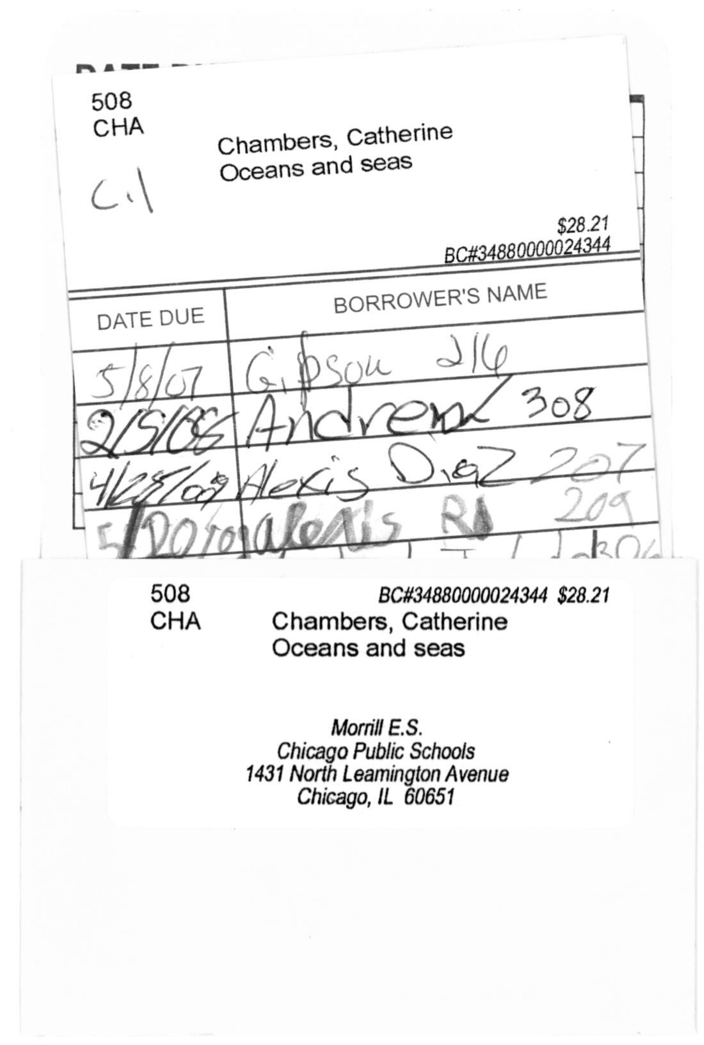
508
CHA
Chambers, Catherine
Oceans and seas
$28.21
BC#34880000024344
DATE DUE
BORROWER'S NAME
508
CHA
BC#34880000024344 $28.21
Chambers, Catherine
Oceans and seas
Morrill E.S.
Chicago Public Schools
1431 North Leamington Avenue
Chicago, IL 60651

Oceans and Seas

Catherine Chambers

Heinemann Library

Published by Heinemann Library,
an imprint of Reed Educational & Professional Publishing,
Chicago, Illinois

Customer Service 888-454-2279

Visit our website at www.heinemannlibrary.com

Designed by David Oakley
Illustrations by Tokay Interactive and AMR
Originated by Dot Gradations
Printed in China

06 05
10 9 8 7 6 5 4 3

Library of Congress Cataloging-in-Publication Data
Chambers, Catherine, 1954-
Oceans and seas / Catherine Chambers.
p. cm. – (Mapping earthforms)
Includes bibliographical references and index.
Summary: Explores the world's seas and oceans, discussing how they were formed, what organisms live there, and how they are used by humans.
ISBN 1-57572-526-6 (lib. bdg.) ISBN 1-4034-0036-9 (pbk. bdg.)
1. Oceans—Juvenile literature. [1. Ocean.] I. Title.

GC21.5.C43 2000
508.162 21—dc21

99-044189

Acknowledgments
The Publishers would like to thank the following for permission to reproduce photographs: P & O, p. 4; Aspect Picture Library/D. Bayes, p. 5; Oxford Scientific Films/C. Bromhall, p. 7; G. R. Roberts, p. 9; Still Pictures/K. Andrews, p. 10; Still Pictures/D. Hinrichson, p. 12; Still Pictures/D. Watts, p. 13; Topham Picture Point/K. Kasahara, p. 14; UK Hydrographic Office (Crown Copyright), p. 15; Oxford Scientific Films/P. Parks, p. 16; Bruce Coleman Limited/J. Burton, p. 18; Oxford Scientific Films/T. Bomford, p. 19; Ecoscene/C. Cooper, p. 20; Still Pictures/F. Dott, p. 21; Still Pictures/B. and C. Alexander, p. 23; Oxford Scientific Films/K. Westerskov, p. 24; Still Pictures/H. Schwarzbach, p. 25; Still Pictures/P. Glendell, p. 26; Ecoscene/P. Fernby, p. 27; Oxford Scientific Films/S. Winer, p. 29.

Cover photograph reproduced with permission of James L. Amos and Still Pictures.

Every effort has been made to contact copyright holders of any material reproduced in this book. Any omissions will be rectified in subsequent printings if notice is given to the publisher.

Some words are shown in bold, **like this.** You can find out what they mean by looking in the glossary.

Contents

What Are Oceans and Seas?

Oceans and seas are mighty bodies of saltwater. They cover nearly three-fourths of the earth's surface. Parts of the edges of oceans are marked by the coasts of huge land masses called continents. Other parts are marked by underwater ridges of rock, islands, or by surges of flowing water called **currents**.

Seas are smaller than oceans and are mostly surrounded by land. They are often connected to oceans by narrow strips of water called **straits**. Some seas are between two strips of land and have large openings at each end. Others lie in scooped-out coastlines. We will see where in the world these oceans and seas lie.

For thousands of years, sailors explored the oceans and seas, looking for trade routes and new lands. The oceans and seas are still used for trade, but now they are used for recreation as well. Ships like the *Royal Princess* take vacationers on cruises all over the world.

How do oceans and sea form?

Oceans and seas formed millions of years ago. Great splits in the earth's crust separated the continents and left huge **basins**. Water ran down mountainsides and hillsides in rivers and streams, and the basins filled with water. As the rivers ran, they picked up salts and other **minerals** from the rocks over which they flowed. The salts and minerals were carried to the sea, as they still are today. We will see how the water in the oceans is changing, and how the ocean floors are moving all the time.

What do oceans and seas look like?

Oceans and seas have many different depths and colors. Their waves can lap the shores gently or rise to huge peaks that batter the coastlines. We will see what causes changes in the appearance of the oceans, seas, and coastlines.

Supporting life

Life on Earth began in the waters of the oceans and seas. Now, they are home to a great range of plant and animal **species**. Humans have lived on the shores for many thousands of years, eating the animals and plants that live in the water. But people have also polluted the water, threatening ocean life. We will find out what the future holds for the oceans and seas of the world.

The oceans and seas provide a rich supply of food. Shellfish are plentiful among the rocks and rock pools of Brittany in northern France.

Oceans and Seas of the World

Counting the oceans

There are three major oceans on the earth. They are the Atlantic, Pacific, and Indian Oceans. Toward the South Pole, these oceans merge together in a **current** of water called the West Wind Drift. This is near Antarctica. Many scientists call these waters the Antarctic, or Southern, Ocean.

Near the North Pole, the Atlantic reaches a circular stretch of water that many scientists call the Arctic Ocean. But some say it is just part of the Atlantic Ocean. The map shows that the Atlantic and Pacific Oceans are divided up into several smaller areas, such as the South Pacific.

Oceans cover different areas of **climate,** as this map shows, but many seas are too small to stretch through different climates. The Mediterranean Sea area, for example, all has about the same climate, which is often called "Mediterranean." Oceans and seas themselves help to form and change the climates more than any other force on Earth.

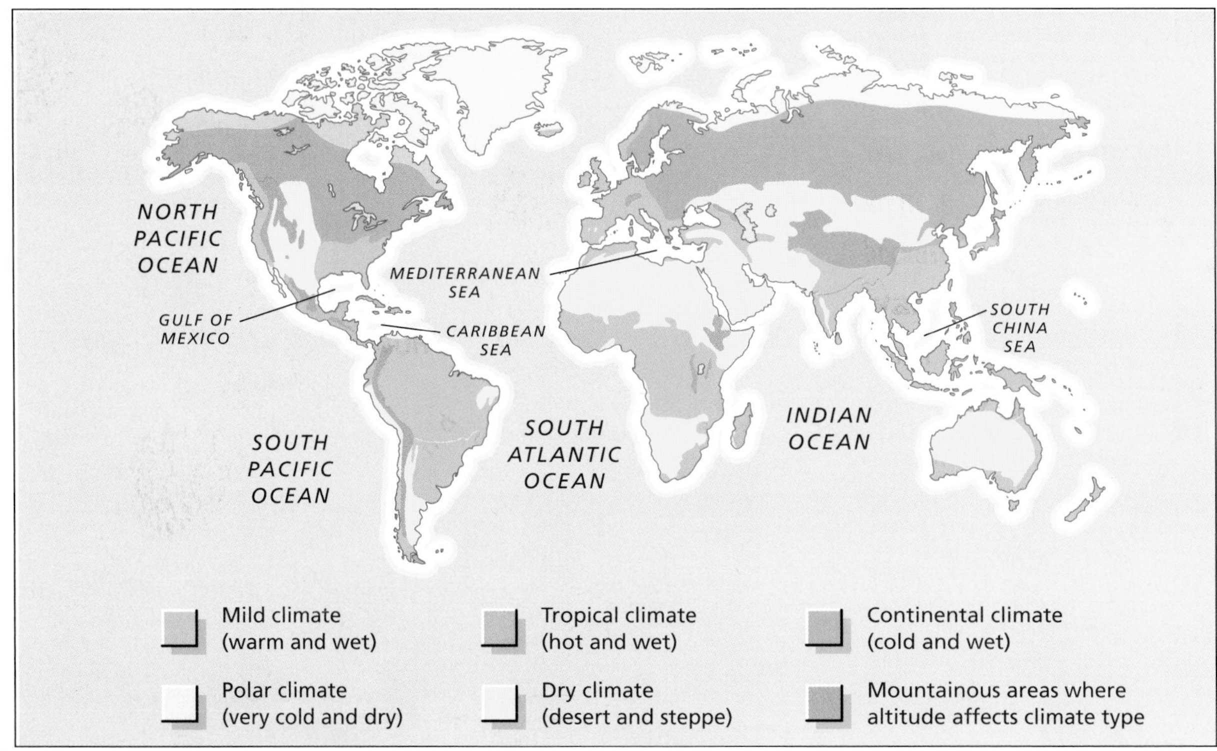

Sorting the seas

Seas are hard to define because they take many different shapes and forms. Sometimes there seems to be no difference between a sea and a gulf or bay, which are scooped inlets near the shore. Sometimes there seems to be no difference between a sea and a channel, which is a strip of water between two coastlines.

Land forms the main borders of most seas. The Caribbean Sea, for example, is surrounded by Central America and a string of islands. The Tasman Sea lies between Australia and New Zealand. Its ends are open, like a channel. The Black Sea is surrounded by land except for a narrow strip of water called the Bosporus.

Sometimes one sea leads into another. This is the Sea of Marmara in Turkey. It lies between the Black Sea to the east and the Aegean Sea to the west.

How Oceans and Seas Form

The earth was once a ball of hot gases. Most scientists think that more than 4 billion years ago, the ball of gases cooled to form rock. There was one continent, called Pangaea. Volcanoes threw out more gases and **water vapor**. This formed clouds that produced rain. The earth's gravity pulled the rain down mountainsides in streams and rivers. The rivers gathered into a huge **basin**, which became a massive ocean. Later, as rivers carried salts and **minerals** to the sea, the water became salty.

Sixty million years ago, the great **plates** of rock on which Pangaea rested were pushed apart by the movement of hot **magma** under the earth's crust. The land split apart, and different oceans and **currents** began to flow around the world.

The earth once had a huge land mass called Pangaea. When it split in two, a warm sea called Tethys separated the northern part, Laurasia, from the southern part, Gondwanaland. Different land masses continued to separate or join together. Finally, Australasia and Antarctica split from each other, which allowed a cold current to circulate around the world.

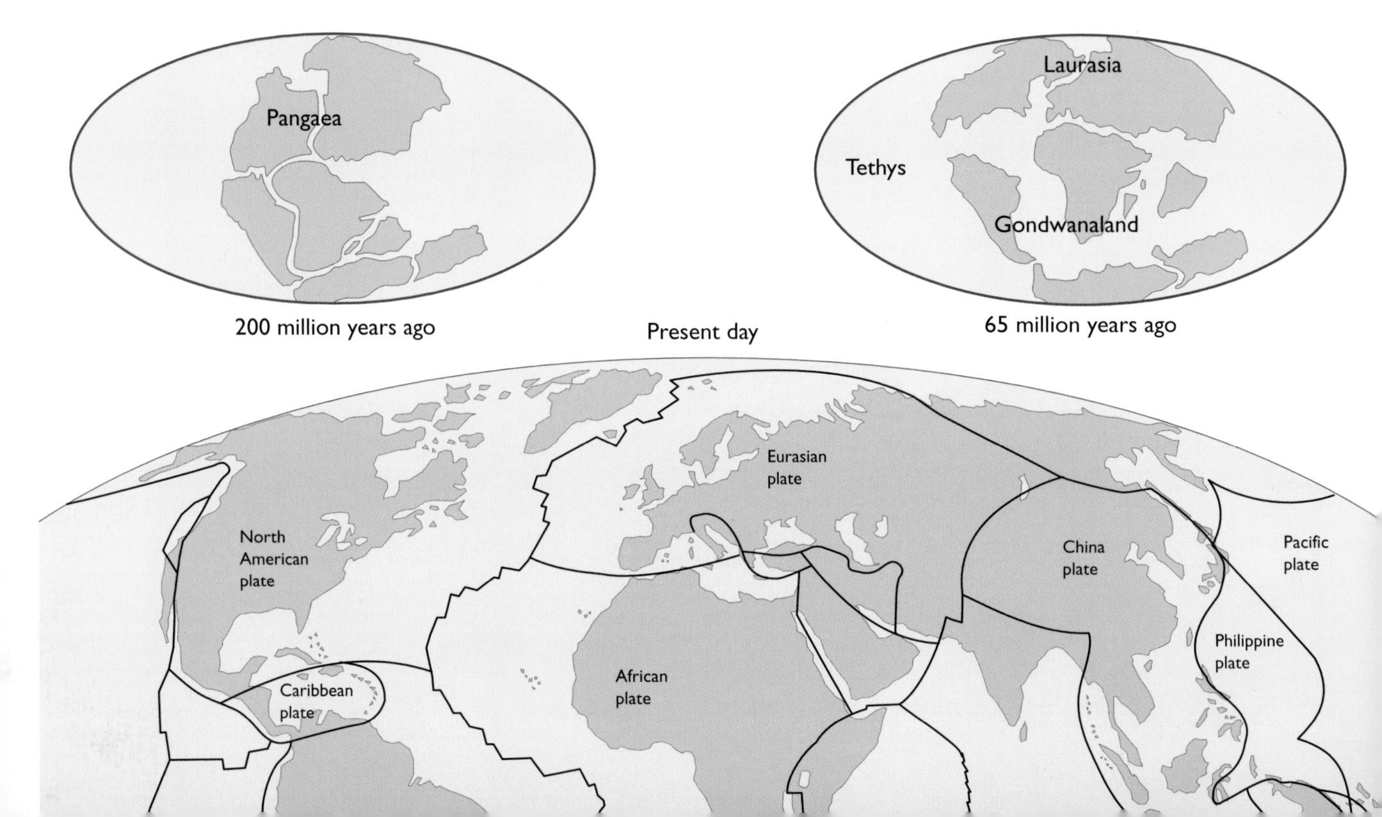

200 million years ago

Present day

65 million years ago

When waves lap the shores, they gradually wear away the cliffs on the coast. This leaves a long, flat **wave-cut platform** that leads to the cliff. The platform is full of rocks, pebbles, and sand. Some of this **debris** is knocked against the cliff, **eroding** it even more. This is known as **abrasion**. The waves cut bays into soft rock, leaving **headlands** sticking out. These get worn away, too. Sometimes a column of rock, called a stack, is left on its own. The stacks in this photograph are called the Twelve Apostles. They are on the coast of Victoria, Australia.

Moving waters

Very salty or cold waters sink, moving the water around and below them. These massive movements of water are called underwater currents. Currents can also be caused by the rotation of the earth or winds blowing in one direction over the water. These winds are called prevailing winds. They cause constant currents, such as the North Atlantic Drift or the Kuro Siwo in the North Pacific. Both of these currents are warm and have a warming affect on the land around them. The gravitational pull of the moon and sun on the water creates the tides, sloshing the water from one side of a sea to another.

Winds blowing from the oceans to the land can bring destructive storms. But in parts of South America, strong winds often blow from the land out to the sea. This causes currents deep in the ocean, which bring water full of **minerals** to the surface. These waters are helpful to sea plants and fish.

Looking Below the Surface

The ocean basin

The ocean **basin** goes down in steps. The steps get deeper as they get further from the continents. The first step is called the **continental shelf**. It runs from the shores of continents and into the oceans for an average of 43 miles (75 kilometers). In some places, though, there is hardly any shelf at all. The waters plunge almost straight down from the continents to a great depth. In other places, the continental shelf stretches out 930 miles (1,500 kilometers).

The second step is called the **continental slope**. It goes down about 8,200 feet (2,500 meters).

We can learn a lot about the earth by studying the layers of **sediment** at the bottoms of the oceans. The ocean floor is covered by a thick layer of sediment. At its deepest point, in the Argentine Basin in the South Atlantic Ocean, it is about 4 miles (7 kilometers) deep. Since 1984, international scientists from the Ocean Drilling Program have used sediments to learn about the history of the earth's **climate**.

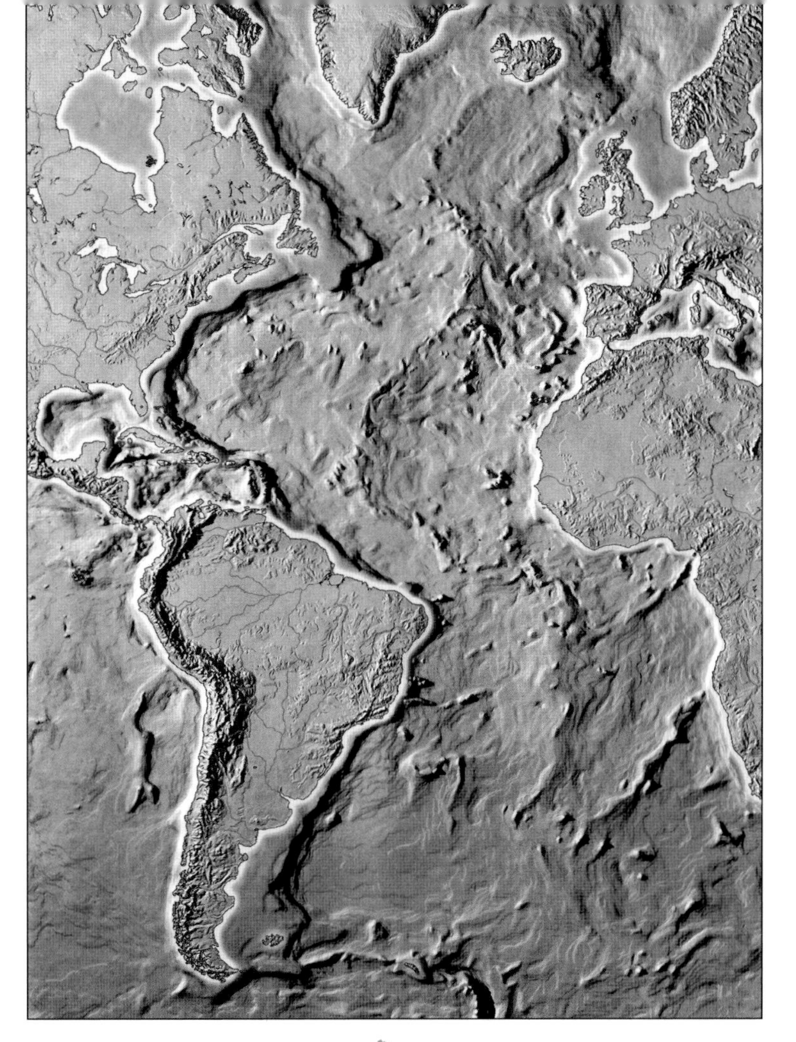

The third part is the continental rise. This is a slope of thick sediment made of rock, soil, **minerals,** and the remains of plants and animals.

Mountain ranges under the ocean are formed when hot, liquid **magma** oozes up from the earth's **mantle** beneath the crust. As it cools, the magma forms new ocean crust, pushing the old aside. This process is called seafloor spreading. Over millions of years, the old crust forms the mountains of the mid-ocean ridges. The Mid-Atlantic Ridge can be seen on this map.

The fourth part of the ocean is a very deep area of flat plains with many mountains. Most mountains lie in chains that form ridges running nearly 4,000 miles (6,500 kilometers) along the ocean floor. Deep trenches plunge from the ridges. The deepest part of the oceans is the Mariana Trench in the Pacific, which plunges 35,840 feet (10,924 meters). Trenches separate the ocean floor into **plates**, which are slowly moving apart. Earthquakes and volcanoes are common in many of the areas where plates meet.

The Importance of Sea Water

Oceans and seas have many **minerals** dissolved in their waters. They have gases, such as oxygen, too. All of these are important for plant and animal life under the waves. But oceans and seas are just as vital for life on land. The water helps to control the **climate**. The water provides rain, too. As the sun shines on oceans and seas, the heat **evaporates** moisture from the surface of the water. This is carried in the air as **water vapor**. Much of the water vapor forms clouds that are blown onto the land. There they drop their moisture as rain.

The different depths of ocean waters are shown by their color. Here, along the Great Barrier Reef off the east coast of Australia, the shallow parts near the shore of the island are a lighter color than the deeper parts. A reef of **coral** in the photograph has pale sea all around it. Coral is made of the shells of billions of tiny sea creatures.

Much of the rain falls on high ground, where it gathers in streams and rivers. This water flows down to the sea, carrying more minerals from the rocks over which they flow. This movement of water from the sea to the land never ends. It is known as the water cycle, or the hydrological cycle. The waters of the oceans and seas also absorb harmful gases from the air. This keeps the gases from rising into the **atmosphere** and polluting the layers of gases around the earth.

In winter in the polar regions, the seas freeze. Icebergs like the one in the photograph break away from huge ice sheets, ice caps, and **glaciers**. Only the tip of an iceberg can be seen. Most of it is under the ocean.

What is sea water like?

There is a lot of salt in the main oceans, and even more salt in some of the seas, such as the Red Sea. There are other minerals, too, such as magnesium, calcium, and potassium. The temperature of the water also varies—from 86°F (30°C) in tropical areas to 29°F (-1.4°C) in the **polar regions**. But however warm or cold it is, the water temperature does not change much throughout the day.

The Mighty Pacific

The Pacific is the oldest, largest, and deepest ocean. Some of its rocks were formed at least 200 million years ago. The ocean contains more than half of the earth's free-moving water. The Pacific also has the deepest point in the world—35,840 feet (10,924 meters) or nearly 7 miles (11 kilometers) below the surface—in the Mariana Trench, east of the Philippines.

The Pacific Ocean's northern boundary is the Bering **Strait**. Its southern boundary is Antarctica. To the east, the ocean reaches the west coasts of North and South America. To the west, the ocean meets Asia, the islands of Malaysia and Indonesia, and the continent of Australia.

This is what can happen when a huge wave called a tsunami shatters the shore. Tsunamis are mighty walls of water that swell up when undersea earthquakes rumble along the ocean floor. An earthquake on one side of the Pacific can cause a tsunami on the other side.

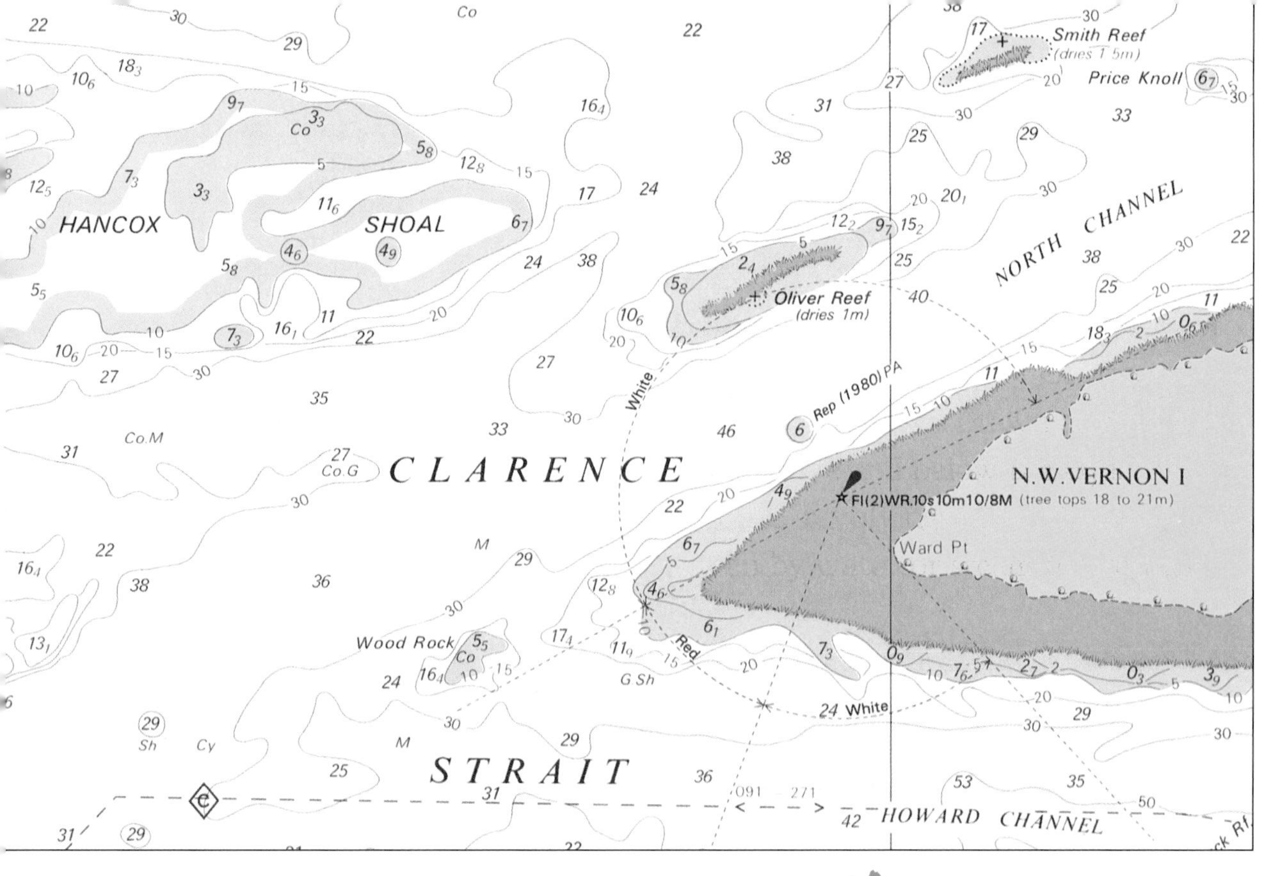

The **continental shelf** is narrow along the coasts of North and South America, but wide near Asia and Australia. The ocean floor is split by wide trenches and studded with chains of mountains.

This map shows part of the Pacific Ocean just off the coast of Australia. This type of map is called a chart. It is used to help navigate ships. The numbers on the sea show the depths of the ocean at low tide. The depths are marked in meters. The chart also shows whether the sea floor is sandy, rocky, or muddy. **Buoys**, lighthouses, and tall buildings along the coast are also marked on the chart.

There are more than 30,000 islands rising from the floor of the Pacific Ocean. The South Pacific area has islands made of **coral**. The western Pacific has a long arc of volcanic islands. This part of the Pacific is called the Ring of Fire because of its volcanoes and earthquakes.

Coral reefs, such as the Great Barrier Reef in Australia, are full of sea life that attracts tourists. In South America, the **currents** have created a good environment for fish. Anchovies are one of the biggest catches. Seabirds also eat the anchovies, and the birds' droppings, called guano, are collected and sold as **fertilizer**.

Plants of the Oceans and Seas

Plants in the water

Green, leafy plants depend on sunlight to make food. In the oceans and seas, these types of plants can only grow where the sun's rays reach into the water. Most of these plants lie in the **continental shelf** area and in the top part of the **continental slope** of the ocean.

The plants are mainly **algae**, which are the oldest form of life able to use sunlight to make food. Algae can range from tiny microscopic plants to fleshy-leaved seaweeds, such as the giant kelp that grows 200 feet (60 meters) a year. Larger algae often have long, thick stalks that attach to the bottom of the ocean. Some attach themselves with clinging roots to rocks near the shore. They have a slimy coating to keep them from drying when the tide is out.

The sargassum weed has given its name to the Sargasso Sea in the Atlantic Ocean. The weed is really a greenish-brown algae that floats freely in the sea and makes the water look green. The Red Sea is full of a red algae, which can range from pink to orange to reddish-brown. There are blue-green and brown algae, too.

The sea buckthorn bush has lots of tightly packed branches and short, thin leaves that do not break in the wind. Its small green flowers have tough brown scales. The flowers and bright orange berries are clustered closely together against the branches for protection. The bush can grow up to 33 feet (10 meters) tall and is found in many parts of Europe.

The smaller plant algae floating in the ocean are also known as **phytoplankton**. They are eaten by tiny creatures called zooplankton, which in turn are eaten by fish. Without algae, there would be very little animal life in the oceans and seas. In 1970 scientists found tiny plant **bacteria** growing on rocks near the hot volcanic vents on the floor of the Pacific Ocean. These bacteria do not need sunlight to make food. Instead, they use a chemical called hydrogen sulphide.

Plants near the shore

Plants near the shore have to cope with strong winds and salty air. Some survive in salt marshes where a river meets the sea. Plants that are able to grow in salty conditions are known as **halophytes**. Many of them get rid of the salt through thousands of tiny holes in their leaves.

Sea Animals

The oceans and seas provide a huge range of **habitats** for animals. These habitats range from the dark, cold depths of the oceans to muddy shores. Like sea plants, most sea animals live in the sunlit area of the **continental shelf** and the upper part of the **continental slope**.

Full of fish

Some fish are bony and have a mouth with jaws. Others have a skeleton made of a soft, flexible material called cartilage, and they have a more flap-like mouth. Fish breathe through gills, which absorb oxygen from the seawater. Fins help fish move through the water. An air bladder inside the body keeps fish from rolling from side to side. Many fish have a protective layer of scales that are covered with a very thin skin. This skin produces an **antiseptic** slime.

The octopus has **adapted** well to life in the ocean. It can grow up to 10 feet (3 meters) across and has eight arms. Each arm has two rows of suction cups, which help the octopus hold its prey. The octopus has a beak-like mouth with poison sacs near it. It uses the poison to stun its catch. The octopus's grayish brown body can change color to match its background. This **camouflage** hides it from enemies. Different species of octopus are found in most oceans and seas.

Some fish have adapted to surviving in the deepest, darkest parts of the ocean. These fish create their own light in their bodies. They do this by producing a chemical called phosphorus.

Many mollusks

Mollusks are a huge family of animals with about 80,000 different **species**. Saltwater mollusks range from oysters, which lie still on the ocean floor, to the busy squid. They all have soft bodies, but most are protected by a hard shell. The space under the shell holds the gills, which are used for breathing and sometimes for collecting food. Small mollusks, like cockles, bury themselves in the mudflats or sand near the shore. They put out a jelly-like feeler called a siphon to suck in water and food.

The porpoise lives in the North Atlantic and Pacific Oceans but prefers the shelter of inlets to the open sea. It is the smallest member of the whale family, which has about 90 different species. Whales are mammals, so they breathe using lungs, but they never go on land. The small porpoise has teeth and eats fish, but some larger whales have no teeth. They strain **plankton** from the water through tiny notches in their mouth bones.

Living with the Ocean

The riches of the sea

Fishing communities have built some of the world's oldest settlements by the oceans and seas. Many of these are next to deep, sheltered bays, which were good stopping places for trading ships. Large ports developed along busy ocean trade routes.

The city of Venice lies between the mouths of Italy's Po and Piave Rivers, which open into the Adriatic Sea. It is protected from the ocean by long barriers of sand, called sandbars. Venice has been a trading post for over 900 years, trading with Far Eastern countries such as China. It is now a modern industrial city, although much of the industry is concentrated on the mainland, around the towns of Mestre and Marghera.

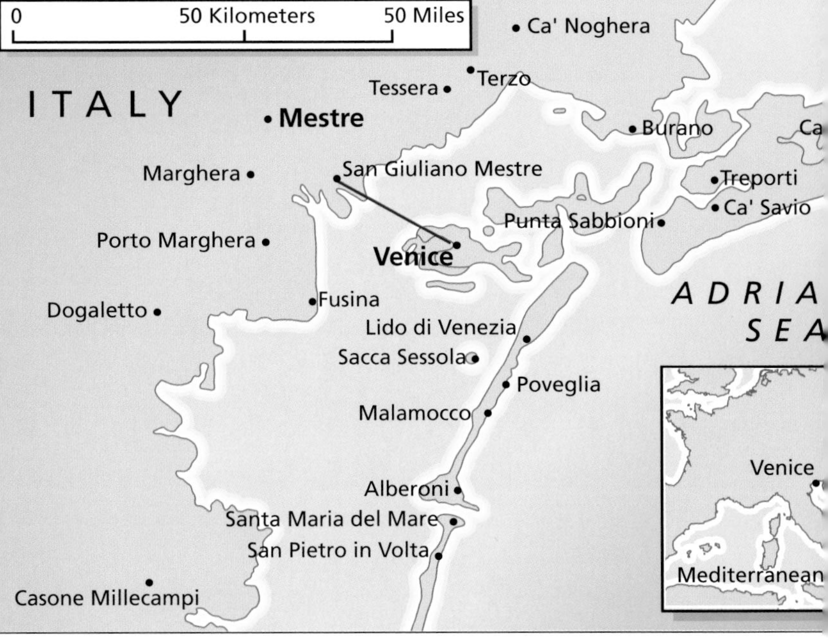

Fishing communities provide the world with an important source of food. Some fishermen still catch fish using traditional boats, such as canoes or wooden sailing vessels. But modern methods allow fish to be caught on a very large scale.

Coastal people also farm or collect seaweed. Seaweed can be used as **fertilizer**, and it is a source of vitamins and **minerals**. The jelly-like substance in brown **algae** is used in products ranging from paints to ice cream.

Sea, sun, and sand have attracted millions of tourists to visit the coasts. In some places, they have changed small fishing villages into huge resorts. People have flocked to the coasts to find work in hotels and restaurants. For some countries, coastal tourism is their most important industry.

Marine minerals

Coastal communities also find work taking minerals from the sea. Salt, magnesium, and bromine are taken from the water. Seventeen percent of the world's petroleum oil comes from the ocean floor. Sand, gravel, and oyster shells are dug from the seabed and used in building materials. Diamonds are also found in some underwater gravel beds.

This is an oil platform off the coast of Norway in northern Europe. Oil has brought many people to live by the sea. It has changed small coastal communities into big industrial towns. The Norwegians have used the money from oil to set up technological industries on the coast. They know that the oil will not last forever.

A Way of Life—Labrador Inuit of Canada

The Labrador Inuit people have lived on the east coast of Canada for centuries. The different Inuit people came to North America from the frozen lands of Siberia in Russia more than 4,000 years ago. They have **adapted** their way of life to the harsh, snow-covered winters and the cool, sunny summers. Today, not all Inuit lead traditional lives. Many live and work in towns and cities.

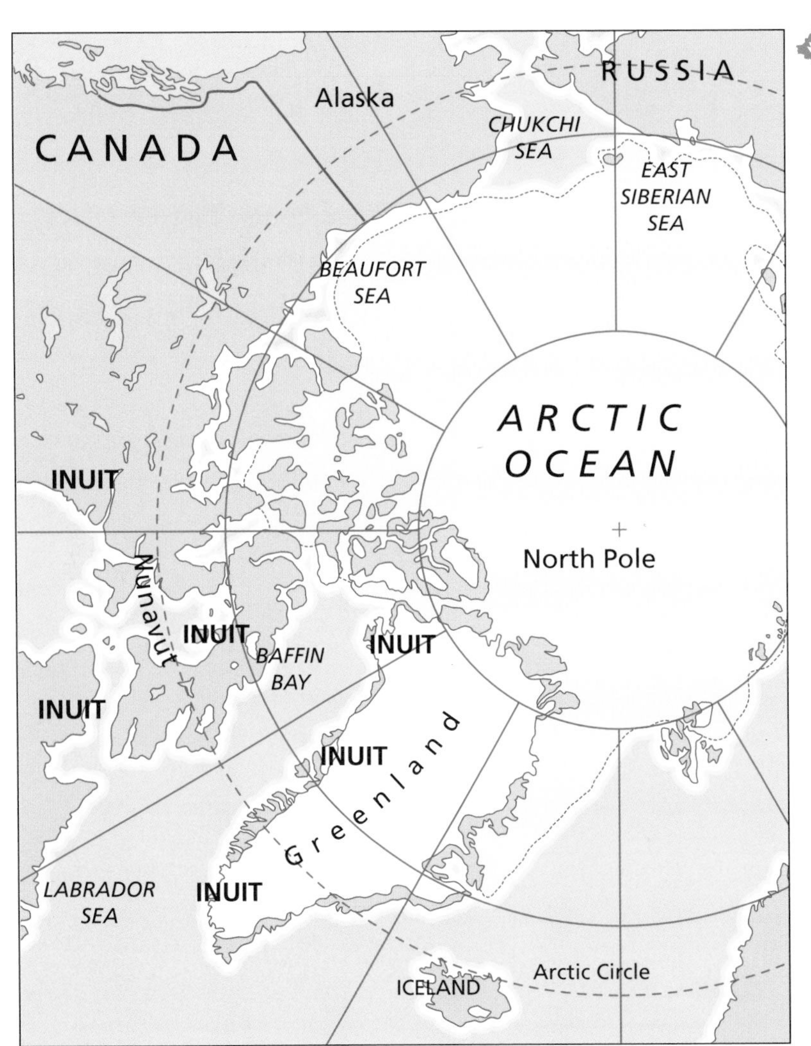

Different groups of Inuit are marked on this map. Every three years, Inuit leaders meet at the Inuit Circumpolar Conference. They discuss issues affecting the region, such as how to preserve the environment. Weather conditions in the Arctic have been changing because of **global warming**. This makes winter hunting difficult. In April 1999, the Canadian government and Inuit leaders created a new territory, covering part of the mainland and many islands. The area has been renamed Nunavut.

Making a living, building a home

Traditional Labrador Inuit use the plants and animals of the coast and sea for their everyday needs. Seals are a good source of food, especially in winter. The Inuit also eat fish, whales, and walruses, which are mostly caught in the summer on the open sea. They preserve any meat not eaten right away by drying or freezing.

The Inuit make seal skins into warm, waterproof clothes. Winter trousers and boots are made with double layers of skin and fur. The traditional coat is a parka, which is a double-layer pullover with a hood. The Inuit also use skins to make **harpoon** lines and to cover tent frames and boats as well.

The Inuit's summer homes are made from walrus or seal skins stretched over a frame of whalebone or driftwood. The walls of winter houses are made of stones packed with moss and earth. Seal fat is burned to make light and heat. Years ago, the Inuit sold carved bone and ivory canes and other goods to traders. Now, to protect wildlife, they use soapstone instead.

For hundreds of years, the Inuit used dogsleds to travel across land. At sea, they used canoes or kayaks, which are made of a whalebone frame covered in sealskin. Some Inuit still use these traditional methods of travel, but others have more modern ways of getting around. These days, many people use snowmobiles to get around in the snow. In the summer, they take their powerboats out to sea.

Sea Changes

Natural changes

Some parts of the world's oceans and seas nearly always have stormy water, such as Cape Horn at the tip of South America. Other parts have long periods of calm, such as the doldrums, which lie in parts of the **Tropics**. But mostly, the ocean waves change all the time. They change with the tides, the weather, and the seasons of the year.

In recent years, some seas have experienced unusually violent seasonal storms. Severe **hurricanes** have battered islands and coastlines. Many scientists believe that they are caused by too much sun on our oceans. This **evaporates** more water, which brings heavier rain and stronger winds.

The amount of fish in the oceans is decreasing. One reason for this is overfishing. At least 60 million tons of fish are caught every year. Many are caught, processed, and frozen on board huge factory ships. A technique known as **sonar** uses sound waves to find huge schools of fish. Lights and electrical pulses attract hordes of fish to the massive nets. When small, unwanted fish are caught, they often die before being thrown back into the sea.

Some scientists think that the **climate** is changing because of flares being thrown out by the sun. Others think the sun is stronger because the earth's protective layer of ozone gases has become thinner.

This looks like chemical pollution on the coastline, but it is perfectly natural. During storms, a jelly-like substance in **algae** gets whipped into a foam. The foam is washed ashore, where it decays.

Many scientists think that **global warming** is caused by the burning of fossil fuels, such as gasoline used in cars and fuels burned in factories and power plants. This releases carbon dioxide and other gases into the **atmosphere**, where they trap the heat from the sun. In the past, most of the harmful gases were absorbed by the oceans, seas, and green plants. But now there is too much pollution to aborb.

What's changing in the water?

Rivers and streams fill our oceans with a continual supply of water and salts. They also carry pollutants. Some are chemical **fertilizers** that have washed into rivers from farmland. Others are **pesticides**. Raw **sewage** and waste from power plants and factories are sometimes pumped directly into the sea. It is also thought that the stronger sun is destroying **plankton** in the water. All of these factors have reduced the numbers of the oceans' plants and animals.

Looking to the Future

The power of the sea

The future of the world's oceans and seas depends on people. We can burn fewer fossil fuels or convert waste gases into cleaner gases. We can use less energy or maybe different kinds of energy. The oceans and seas can actually help us to do this. Wave and tide power close to the shore can turn **turbines** to make electricity. One recent idea is to use the ocean's thermal energy. This energy comes from the sun's heat that has been absorbed by the water. It comes from ocean **currents**, too. This solar heat can be changed into electric energy in a process called Ocean Thermal Energy Conversion.

This oil spill came from a huge oil tanker, the *Sea Empress*, which broke up near Milford Haven in Wales. The oil killed thousands of birds, fish, and shellfish. It covered the sands and spoiled an area known for its natural beauty and wildlife. Many governments have passed safety laws to try to prevent these kinds of accidents in the future.

Fish forever

The number of fish in the seas is decreasing. More than 90 percent of the fish caught in the world comes from oceans and seas. Many small fish are destroyed before they can become adults and produce more fish. We can stop catching small fish by using nets with larger meshes. We can also develop more fish farms along the coasts, so that fewer fish from the sea are caught.

There are some things that we cannot help. In the 1970s, the number of anchovies off the coast of Peru began to fall dramatically. Fishermen were blamed for catching too many. But it is now known that this was only partly true. Many schools of anchovies were moving away from the cold Peruvian Current to find warmer waters.

We can also help ocean life by cleaning the waste from our factories and power stations before it flows into the sea. We can increase the amount of freshwater that reaches the sea by building fewer dams across rivers. The dams are keeping water and the **minerals** in it from reaching the oceans.

Some scientists believe that **global warming** is melting ice in the **polar region**. The meltwater raises the level of seawater. This extra water and increasingly violent storms are quickly **eroding** the coastlines. In some parts of the world, towns and villages are falling over cliffs and into the sea. Coastal defenses, like these in Norfolk, England, can protect the coastlines.

Ocean and Sea Facts

Top twelve seas

These are the top twelve seas in order of size. Seas do not have exact boundaries, so sizes are estimated. The column on the right shows the seas' average depths. The table also shows to which ocean the seas are connected.

Sea	Location	Area (sq. mi.)	Area (sq km)	Average depth (feet)	Average depth (m)
South China Sea	Pacific	1,148,196	2,974,600	4805	1464
Caribbean Sea	Atlantic	1,067,676	2,766,000	8450	2575
Mediterranean Sea	Atlantic	971,176	2,516,000	4926	1501
Bering Sea	Pacific	875,448	2,268,000	4893	1491
Gulf of Mexico	Atlantic	595,598	1,543,000	5300	1615
Sea of Okhotsk	Pacific	589,808	1,528,000	3193	973
East China Sea	Pacific	482,144	1,249,000	620	189
Hudson Bay	Atlantic	475,552	1,232,000	305	93
Sea of Japan	Pacific	389,008	1,008,000	5471	1667
North Sea	Atlantic	221,950	575,000	308	94
Black Sea	Atlantic	178,332	462,000	3908	1191
Red Sea	Indian	169,068	438,000	1766	538

Plunging the depths

Deepest ocean—Pacific Ocean, with an average depth of 13,200 feet (4030 meters).

Next deepest—Indian Ocean, with an average depth of 12,800 feet (3,900 meters).

Tsunamis are huge waves caused by undersea earthquakes and volcanic eruptions. One of the most destructive tsunamis happened in 1755 in the Atlantic Ocean. It hit the port city of Lisbon, Portugal.

The highest wave ever recorded was 112 feet (34 meters) high. This surfer is enjoying a ride on a wave off the coast of Hawaii. Hawaii has some of the biggest waves in the world.

Glossary

abrasion erosion caused by moving stones carried by wind or water

adapt to change and make suitable for a new use

algae simple form of plant life, ranging from a single cell to huge seaweed

antiseptic having the ability to kill germs

atmosphere layer of gases that surrounds the earth

bacteria tiny, one-celled organisms

basin sea or ocean floor that slopes downward like the inside of a bowl

buoy anchored float used to mark a channel or obstruction in the water

camouflage way of coloring or covering something so that it is difficult to see against the things around it

climate rainfall, temperature, and winds that normally affect a large area

continental shelf relatively shallow area that slopes from the coast of a continent into the ocean

continental slope area under the ocean that runs from the edge of the continental shelf further into the ocean to a depth of about 8,200 feet (2,500 meters)

coral hard rock made of the shells of tiny, dead sea animals cemented together with limestone

current strong surge of water that flows constantly in one direction

debris eroded material, such as rocks, pebbles, and sand

erosion wearing away of rock or soil by wind, water, ice, or acid

evaporate to turn from liquid into vapor, such as when water becomes water vapor

fertilizer substance added to soil to help plants grow better

glacier thick mass of ice, formed from compressed snow, that flows downhill

global warming gradual increase in the earth's temperature

habitat place where a plant or animal grows or lives in nature

halophyte plant that can grow in salty soil

harpoon spear, attached to a line, that can be shot from a special gun to catch whales and fish.

headland cliff sticking out into the sea at the end of a bay

hurricane severe storm with wind that blows faster than 75 miles (120 kilometers) per hour

magma hot, melted rock beneath the hard crust of the earth

mantle layer of hot, molten rock on which the earth's crust rests

mineral substance formed naturally in rock or earth, such as oil or salt

pesticide chemical used to kill insects

phytoplankton tiny algae

plankton tiny plants or animals in the water

plate area of the earth's crust separated from other plates by deep cracks. Earthquakes, volcanic activity, and the forming of mountains take place where these plates meet.

polar region area around the North or South Pole

sediment fine soil and gravel that is carried in water

sewage human waste material

sonar device that measures distances and depths by sending sound waves that hit an object and echo back

species one of the groups used for classifying animals. The members of each species are very similar.

strait narrow strip of water connecting a sea to an ocean

trade winds winds that blow steadily towards the equator, but get pulled westward by the rotation of the earth

Tropics region between the Tropic of Cancer and the Tropic of Capricorn, two imaginary lines drawn around the earth above and below the equator

turbine revolving motor that is driven by water or steam to produce electricity

water vapor water that has been heated until it forms a gas, which is held in the air. Drops of water form again when the vapor is cooled.

wave-cut platform long, flat step of eroded rock, pebbles, and sand that leads to a coastal cliff

More Books to Read

Clarke, Penny. *Beneath the Oceans.* Danbury, Conn.: Franklin Watts, Inc., 1997.

Kovacs, Deborah. *Off to Sea: An Inside Look at a Research Cruise.* Austin, Tex.: Raintree Steck-Vaughn, 2000.

Oldershaw, Callie. *Oceans.* Mahwah, N.J.: Troll Communications, L.L.C., 1997.

Riley, Peter. *Our Mysterious Ocean.* Pleasantville, N.Y.: Reader's Digest Children's Publishing, Inc., 1998.

Smart, Brian Hunter. *Oceans.* Brookfield, Conn.: Millbrook Press, Inc., 1999.

Index